The Physical and Mental Differences of Humanity

Daniel Pickford-Gordon

The Physical and Mental Differences of Humanity © 2009
Daniel Pickford-Gordon

ISBN 978-0-9561601-0-2

All rights reserved. No part of this publication may be reproduced, stored in a retrieval system, or transmitted, in any form or by any means, electronic, mechanical, photocopying, recording or otherwise, without the prior permission of the publisher.

Contents

Introduction	4
Part 1: The Differences of Humanity	7
The Implications of Language	8
The Implications of History	9
The Changeable Physical Differences	10
The Unchangeable Physical Differences	12
The Unchangeable Mental Differences	14
Part 2: The Levels of Humanity	20
Level A	22
Level B	25
Level DE	27
Level C	30
Level F	33
Level G	36
Level H	38
Level IJK	39
Level L	42
Level M	43
Level NO	44
Level P	48
Level S	50
Level T	51
Conclusion	52

Introduction

The races of humanity differ from each other both physically and mentally. This is a fact which will not change: tears and screams generated in order to alter this truth shall fall upon deaf ears. It thereby falls upon me to correctly define the word "race", which I shall now endeavour to do.

Race has been described in a variety of ways throughout history. People speak of whites, blacks, and others; yet they also speak of the British, the French, the Ethiopians, and the Egyptians; and sometimes they speak of a number of systems of classification. Although the real system shall be revealed, the unveiling of the classification scheme is not the sole purpose of this book. Part 1 details the variations of the human spectrum in five sections: the first section describes the implications of language; the second section describes the implications of history; the third section describes the changeable physical differences of human beings; the fourth section describes the unchangeable physical differences of human beings; and the fifth section describes the unchangeable mental differences of human beings. Part 2 details the races themselves. The conclusion immediately follows Part 2.

The races of humanity can be thought of as levels of humanity, with each level corresponding to a different human evolutionary stage. A current myth is that human beings ceased to evolve a considerable time ago, and that all human beings currently in existence are identical. It is a truth that one level is superior or inferior to another level, hence usage of the word "level". Evolution is, always has been, and always will be, a gradual process.

I would also like to state that if an individual fails at this task or at that task, it is not necessarily the case that the individual is lacking in intelligence, or that the individual is poorly evolved. Also, if an individual is at a lower level and, sometimes solely in theory, is incapable of reaching the peaks which an individual at a higher level can reach, this fact does not make that individual "worse" than the other. It is frequently the case, in my view, that individuals of lower races can be "better" than individuals of higher races. I would also like to point out that primitive individuals within humanity are nowhere near as primitive as they appear to the eye.

Written sources, photographic sources, auditory sources, video sources, and memories of individuals I have either known, encountered, or observed by chance, were all utilised in order to write this book. I am thankful for the website "Wikipedia", which has helped me to learn about the "Y Chromosome DNA Haplogroups" and the "MTDNA Haplogroups". I am also thankful for the "Y Chromosome Consortium", responsible for the naming of the "Y Chromosome DNA Haplogroups", and its members, whoever they may be. I would also like to thank the individual or individuals responsible for naming the "MTDNA

Haplogroups". Without "Wikipedia", "Y Chromosome DNA Haplogroups", and "MTDNA Haplogroups", this book would not have been composed in this form, and without "Wikipedia" and "Y Chromosome Haplogroups"-although a similar though woefully reduced book could have been created using solely "Wikipedia" and "MTDNA Haplogroups"-it is likely that I may never have found the answer. I am, and shall always be, grateful for their aid.

Part 1: The Differences of Humanity

In this section the multitude of ways in which humanity differs shall be described. In essence this section is an explanation of the standards by which the races shall be judged and labelled in Part 2. I would also like to stress that the possession of primitive features in many cases does not equate to primitivity. However, this does not change the fact that the feature or features possessed by the individual are, indeed, primitive. While the features described as primitive are most obvious in extinct hominids and other lower primates, they are nonetheless subtly present in humanity, to varying degrees.

The Implications of Language

Language is the method of human communication which consists of the usage of words in a structured and conventional way. If the users of a language have adopted a method of communication which is simple, limited, and poorly evolved, then this can reflect a certain fact about its users. However, if one language is more complicated than another language, this does not imply that the language is superior: complications could be completely unnecessary. Much has been lost concerning the subject of language, and a complete reconstruction may be impossible. It is also the case that the problems caused by natural hazards must be taken into account.

The Implications of History

History is the study of past events. The histories of various cultures and societies shall be discussed within the main body of text. While creation and preservation of aspects of history can show us that a race is highly evolved, lack of this aspect does not necessarily condemn a race to primitivity. As in the case of language, much has been lost concerning the subject of history, and a complete reconstruction may be impossible. It is also the case that the problems caused by natural hazards must be taken into account.

The Changeable Physical Differences

There are many physical differences within humanity which are changeable through sexual reproduction. The features can be gained, and they can be lost, to varying extents. I shall now describe these features, which shall be covered more extensively in Part 2. There are a wealth of physical features that can be changed which shall not be described, or which shall be merely touched upon, for instance, aspects of hair, eye colour, and height. All changeable physical features are quantitative as opposed to qualitative, changing in differing degrees through sexual reproduction as opposed to immutably being passed on from father to son, or from mother to daughter.

    Tone of voice is a feature associated with primitivity. A deep, toneless voice is a feature of the primitive, while a light, toneful voice-capable of much variation-is a feature of the advanced. This is an indication of our evolution from lower primates, since features of their voices can be compared to the deepest and most toneless voices of humankind. However, this feature is changeable, and can be passed on through reproduction. The most advanced individuals can be born with this manner of voice: possession of this feature of voice is not a condemnation, merely a genetic indication. Individuals possessing voices which have deep and toneless qualities about them are sometimes frowned upon in society. However, even if individuals possess this manner of voice, the variation that this manner of voice is capable of can be taken advantage of to great extent.

Aspects of the hands and the feet is another feature associated with primitivity. Long hands and feet shaped in a certain manner are characteristic of the primitive ancestors of human beings.

The thickness of the nose is a quality that can be passed on to an individual's generations. Although the thickness and flattened aspect of an individual's nose is a primitive feature which can be passed on through reproduction, there are other primitive aspects of the nose which can never be passed on through reproduction.

Thickness of the lips is another changeable feature. It can be passed on through reproduction. Lips which are large and thick represent primitivity, while lips which are thin and small represent advancement.

The protrusion of the ears-with the ears lying at an angle to the skull as opposed to the ears lying flat against the skull-is a changeable feature associated with primitivity. The protrusion of the ears at sharp angles from the skull represents primitivity, while the flattening of ears against the skull represents advancement.

The Unchangeable Physical Differences

There are many physical differences within humanity which are unchangeable through sexual reproduction. I shall now describe these features, which shall be covered more extensively in Part 2. There are physical features that cannot be changed which shall not be described, for instance the distinction between being either male, or being female. Others shall merely be touched upon. All unchangeable physical features are qualitative as opposed to quantitative, being passed on directly from father to son, or from mother to daughter.

Patterns of the voice is an unchangeable feature. Primitive voices have a distinct pattern and feel to them, while there is no such thing with the voices of the advanced.

Aspects of the human walk are also unchangeable. The walk has evolved from a general mockery and exaggeration into a spectacle which can be graceful.

The formation of the nose is also unchangeable. Primitive noses are poorly formed, and lack the contours of the noses of the highly evolved.

Protrusion of the lips and the teeth is another feature which is unchangeable. In the primitive, the lips and the teeth protrude, and the lower half of the skull appears, in profile, to be triangular; in the advanced there is no such protrustion, and the lower half of the skull appears, in profile, not to be triangular, to any extent.

A domed skull is another feature which is unchangeable. In the primitive, the skull appears rounded, while in the advanced the skull has more horizontal and vertical aspects.

A sloping forehead is another feature which is unchangeable. In the primitive, the forehead slopes downwards at a slight angle, while in the advanced the forehead does not.

Protrusion of the lower jaw in a distinct manner is another unchangeable feature. In the primitive who have not developed in the East, the lower jaw protrudes distinctly, while in the advanced who have not developed in the East, it does not, or protrudes only slightly. Alternatively, in the primitive who have developed in the East, the distance from the upper lip to the nose is large, while in the advanced who have developed in the East, it is small.

The blending of the cheeks with the jaw area is another feature which is unchangeable. In general, the primitive have cheeks which appear separate from the jaw area, with a circular area appearing around the lips which blends poorly with the cheek area, while in the advanced there is no such thing, and the area around the lips blends smoothly with the cheek area.

The Unchangeable Mental Differences

There are many mental differences within humanity that are unchangeable through sexual reproduction. These features are qualitative as opposed to quantitative, being passed on directly from father to son, or from mother to daughter. I shall now describe these features, which shall be covered more extensively in Part 2. There are mental differences that cannot be changed which shall not be described, or which shall be merely touched upon.

I shall now briefly describe the concept of change. Individuals can change in several ways: for instance, changing to become suited to the environment, changing the environment to become suited to themselves, or by finding a new environment. Individuals who are incapable of performing such or similar tasks can be said to be unintelligent, and poorly evolved. They are primitive, in ways, and are low in the racial hierarchy. The issue of problem-solving is closely tied to this. A primitive individual will repeatedly fail to solve problems, sometimes using the same unsuccessful methods ad infinitum, while an advanced individual will succeed. Also connected is the idea of learning. Primitive individuals frequently fail to learn from mistakes, while advanced individuals frequently learn from mistakes.

Complexity is another issue by which individuals can be judged. A simple individual is poorly evolved, while a complicated individual is highly evolved.

Creativity is a further issue by which individuals can be judged. The most evolved create a multitude of wonders, while the least evolved do not.

Judgement is another issue which relates to primitivity, evolution, and intelligence. Mistimings and misjudgements are characteristic of inferior individuals, while good timings and good judgements are characteristic of the highly advanced.

Reasoning is associated with race. There are a multitude of issues the poorly evolved fail to understand, while the well evolved succeed in the understanding of a multitude of issues.

Curiosity is also connected to the idea of the existence of stages within the evolutionary development of humankind. The primitive are blissful in ignorance, and do not desire answers, while the superior question, and seek to know.

Memory is also linked with evolution. The primitive frequently struggle to remember, while the advanced frequently excel at remembering.

Perception is a further standard by which individuals can be judged. The inferior frequently fail to perceive well, while the advanced frequently excel at perception.

Purpose is another standard by which individuals can be judged. The primitive frequently equate purpose solely with sexual intercourse, sexual reproduction, food and feeding, and other activities of such nature, while the advanced equate purpose with more complicated activities, sometimes ceasing to engage in basic activities in order to involve themselves with more complicated activities. Examples of complicated activities include working towards inventions, innovations, discoveries, creations, cures, developments, and improvements.

Aggression is a lack of control, and a primitive aspect; displaying or to possessing aggression is frequently equated with displaying or possessing weakness, and being uncivilised. Aggression is one of the more noticeable ways in which higher races are separated from lower races. The lower races are highly aggressive, while the higher races are much less aggressive. In addition to "turning to" aggression frequently-having aggressive tendencies-the poorly evolved also express their aggression in less complicated, less pleasant, and less polite ways than the highly evolved, and their expressions of aggression are frequently very blunt indeed. Violence and profanity would be found more often among the lower races, if not for the existence of radically different cultures, customs, traditions, and the like possessed by the various countries, as well as several other factors.

Emotions inform the brain of various events, processes, and issues, among others. Therefore if certain emotions are not possessed by an individual, or if certain mixtures of emotion, intensities of emotion, or subtleties of emotion will never be present in an individual, then that means that the emotions present, which are primitive, fail to inform the brain correctly; interpretation of oneself and interpretation of the external is flawed. For instance, if an individual lacks fear, that individual will not flee from danger as well or as quickly as the individual ought to. Pity is lacking in lower races. Empathy is one of the ways in which Levels R, S, and T excel over individuals of lower levels; this is much in evidence in the superiority of Level R, S, and T actors and actresses, counsellors, and interpreters of art, among others, over lower levels within humanity. It is not possible for one to properly form a connection with an individual at a lower level to oneself. Emotions are linked to judgement and perception, which concern intelligence. Emotions are poorly developed in the lower races, sometimes appearing obvious and limited.

    Emotional complexity is an issue related to evolution. Certain mixtures of different emotions are lacking in the lower races, while in the higher races various mixtures of various emotions are present.

    Emotional intensity is another concept concerning race. The primitive are incapable of being consumed with great emotion and also cannot develop or display small amounts of emotion, while the advanced are frequently consumed with great emotion, and frequently develop small amounts of emotion. It can be said that there is no subtlety about the primitive.

The ability to display emotion about the eyes and the lips is one more factor concerning primitivity. If one is unable to display emotion correctly, one cannot connect well with others as one should; visual display of emotion is an aid, and helps to eliminate confusion. The lower races cannot display emotion well in and around the eyes, and around the lips, while the eyes of the higher races are full of emotion, and the lips change in certain ways to display emotion well.

Finally, looks are also an indication, but solely an indication, of whether an individual belongs to a different level of humanity, and the attraction of individuals to members of their own level and not to members of other levels is an indication of this. There is always a sense of "other" when conversing with individuals of other levels. However, there are differences. When individuals belonging to mine own Level R observe and converse with individuals of low levels, they sometimes experience varying degrees of revulsion, or find the individuals sinister. However, pure, ominous fear is never felt when observing and conversing with these individuals. When individuals belonging to mine own Level R observe and converse with individuals of Levels S and T this feeling is sometimes present. They are indeed "higher beings", and the only real humans, although one should acquaint oneself with the genetic makeup of their recent ancestors, whatever it may be, before one falls to one's knees in worship. There is indeed more to an individual than the level they have attained.

Part 2: The Levels of Humanity

It is now a necessary to define the word "race". Race roughly corresponds to the groupings of "Y Chromosome DNA Haplogroups" in men and to the groupings of "MTDNA Haplogroups" in women. These are indications of the levels of humanity, and they do not correspond well. Indeed, the greater number of haplogroups corresponding to women can mean that there are more races of women than there are of men; and frequently the "female equivalents" to the male individuals of a level are not found alongside them, either in anywhere near the same numbers, or sometimes, even at all. Because of this complication I shall describe the male levels of humanity, and shall list their female equivalents alongside them in certain cases. It is fact that an individual is not solely male or female, but is a male of this or that level, or a female of this or that level. The factor of being of a level is passed on exclusively from father to son, and from mother to daughter, and it can never be changed; it is part of what it means to be male or female; it is part of the "package". The levels evolved some considerable time ago, when the climate of the world was very different, as was the terrain. Once the levels had been formed, had "solidified", so to speak, there was finality. It is frequently the case in life that once an object is formed, it cannot be unformed, except if it is destroyed, and this is true of the levels. It must also be noted that, due to the incredible amounts of variation within groups E, J, O, and R, and their great geographic spread, that certain physical features of individuals belonging to these groups often mimic physical features of individuals belonging to other groups, even if the aforementioned individuals are not of their type, but are instead of E type, of J type, of O type, or of R type. This is an example of a form of evolution which has been observed readily in nature- convergent evolution. While it is true that individuals of certain levels are equal mentally, physically they are "Eastern", "Western", or other versions of each other, with the original features being stretched or squashed accordingly. The male levels are named after the "Y Chromosome DNA Haplogroups", while the female levels are named after the "MTDNA Haplogroups". When geographic distributions have been described, migrations of the recent past have been ignored. "Level A" may have appeared around one-hundred-and-seventy thousand years ago, and ten thousand years ago all levels had evolved. There are a large number of groupings within humanity, because there are a large number of races within humanity.

Level A

The female equivalents are designated "L", for instance "L0", and "L1". Individuals of this level can be observed among the San of Africa.

The native language is made up of clicks; it is called Khoisan. This system is highly flawed, for, among other things, a click is toneless, and cannot convey the complexity of languages that involve the voice as such; there is, indeed, only so much a click can do. The language seems to be more for convenience rather than being an organised system of communication, and appears to be frequently used in order to explain certain relatively simple concepts. There are several examples: the Khoisan for "bridge" is "hole upper side", which displays a lack of understanding of the function of the bridge, among other things; and in Khoisan there is no decimal system: only numbers one to ten are used.

History reflects facts about the primitivity of Level A individuals; development has been woefully inadequate, and many say it has been basically non-existent.

Tone of voice is deep, and the hands and feet are long. The nose is thick, and flattened. The lips are extremely thick, in particular the upper lip, more or less uniquely to Levels A and B, which behaves as though it desires to blend with the area of skin above it, a curious feature. Generally, the ears noticeably "stick out" from the skull, as opposed to lying flat in the manner of the ears of higher levels.

    The voice pattern is characteristic of Levels A, B, DE(although this natural pattern is often disguised by individuals of Level DE, as shall be explained), and, to a lesser extent, C; phrases are as short as possible, and are broken up, giving a halting impression to speech-it appears "chopped up". The walk is exaggerated, with the same primitive "bouncing" quality which can be found in Levels A, B, DE to a lesser extent, and faintly in C. It may represent a difficulty in evolving to bipedalism. The nose has been formed in what can be described as a pitiful manner: there appears to be no line of cartilage present at all between the tip of the nose and the space between the eyebrows; the nostrils are poorly formed, and lack the shape of the nostrils of higher races; and the sides of the nose by the nostrils are ill formed, making the nose tip appear more fleshy and vague than the nose tip of individuals of higher levels. The teeth protrude noticeably and significantly. The skull is heavily and infamously domed, being almost of a spherical shape. The cheek area and the jaw area blend poorly, creating a distinctive appearance. The "form" of individuals of this level-a dragonfly-like appearance, with the eyes far apart and large-is location-specific, since the "form" evolved at a specific location in Africa. This form is often emulated by individuals of Level E, due to the spread of Level E individuals across Africa and also into other regions. There, however, many similarities end.

It has been said that individuals of this level have more or less failed to change. The San are not complex, in comparison to the other levels of humanity. They are also lacking in many other areas associated with the mind, and the lips indeed fail to display any emotion. The San are the most aggressive and hostile of all races of humanity. The form of the San is not a form designed to display emotion well; if the form were different, they would appear even more hostile than Level B individuals. The eyes do however appear dead, it has often been said, but there is insufficient aggression displayed by the eyes because of the form. Many individuals are extremely adverse to the San, as they are to Level B individuals, and frequently fail to contain their revulsion. Level A individuals, like Level B individuals, are often converted into slaves, or are eaten, by the Level E individuals which dominate Africa.

Level B

The female equivalents are designated "L", for instance "L0", and "L1". Individuals of this level can be observed among the Pygmies of Africa.

The Pygmies appear to have lost their native language; this is most unfortunate, as it would doubtless form a link between the native language of Level A individuals and the native language of Level DE individuals.

History of the Pygmies, as in the case of Level A individuals, appears to be lacking, because there does not seem to be much of it. In ways, like Level A individuals, they have failed to develop.

Tone of voice is noticeably deep, and hand and foot are of considerable length. The nose is both thick and flattened, and lips are thick, in particular the upper lip. The ears stick out.

Phrases are broken up. The voice, unlike the voice of Level A individuals, is full of aggression and hostility, which reflects the hostility ever-present in the eyes. The walk is exaggerated, and the "bounce" of the walk is omnipresent. The legs move forward in a manner akin to the harvestmen of the phylum Arthropoda, with each leg sticking out as far as it will go, away from the body and forward, giving them a spider-like walk. The nose subtly begins to take on a vague form. Lips and teeth protrude, and the skull is domed, though it is not as domed as the skull of Level A individuals. The forehead slopes noticeably. The lower jaw infamously and distinctly protrudes, and again cheek and jaw area blend badly, creating a "Level B patch", similar to the "Level A patch". This form is, again, a form that evolved in a specific region of Africa, a different region to that of Level A, and again the form is emulated by some Level E individuals. Here, instead of primitivity being displayed noticeably in roundness of skull and deadness of eye, the form is designed to display primitivity in the sloping of forehead and in the protrusion of jaw, as well as in the obvious hostility of eye.

      Like Level A individuals, Level B individuals have not come to be associated with change, or complexity, and lack in many areas of the mind, including emotion, and the lips fail to display feeling. They display hostility and aggression in the eyes but there is also a sense of emptiness there. Like Level A individuals, they are frequently converted into slaves, or eaten, by Level E individuals.

Level DE

Level DE consists of Level D and of Level E, which are equal mentally, with Level D being an "Eastern" or "Asian" version of the African Level E. Although Level DE evolved near the northeast of Africa, Level D "went east" while Level E spread in different directions, and the DE form has been altered accordingly. Level D is found among the Ainu of Japan, and the Tibetans of China. Level E is found among most African peoples, although within Africa there are many individuals of other levels present. Levels A and B appear to resemble each other more than Levels B and DE, and the developmental "jump" or "gap" between Level B and Level DE may be the same length as the developmental "jump" or "gap" between Level DE and Level C.

    The history of Level D and Level E individuals appears to be relatively unimpressive, although what they have achieved shows their superiority to Levels A and B.

    Voice tone is deep, hands and feet are long, the nose is thick and lies close to the skull, the lips are thick(yet noticeable thinner than the lips of Levels A and B, lacking the enormity of upper lip), and the ears do protrude.

    The voice pattern is broken up-to an extent. Many individuals of Levels D and E display good reactions when conversing with others, and appear to form unbroken conversations and monologues of some fluidity on numerous occasions. However, many have remarked that not much is said during these conversations and monologues; the Level D or Level E individual may appear to speak, but what is actually said appears to be lacking in scope and such like; many say that the individual merely attempts to delay the pace of the conversation in order for the brain, which does not function at the speed of the brain of a higher level individual, to process the information and form some kind of a convincing answer or statement. Walk has evolved considerably, although there is always a "bouncing" aspect to the walk-an aspect which has not yet been dispensed with. However, this aspect is nowhere near as stereotypical as this aspect within A and B. The nose begins to take on a certain formation, although the nostrils are still ill formed, the line of cartilage from the tip of the nose to eyebrow level is still rather vague, and the area around the nostrils by the sides of the nose is always ill formed and vague. For Level E, the lips and teeth seem to protrude, and the skull is diagnostically domed somewhat, although this is often difficult to observe. For Level E, the forehead slopes, diagnostically, and a distinctive characteristic of every Level E individual is a protrusion of the lower jaw area. For Level D individuals the distance between upper lip and nose is always large. In spite of this, it must be remarked that Level D is still a Level DE "type", and the eyes and aspects of skin, face, and lip are never as "broad" as the eyes and aspects of skin, face, and lip of Levels C and O. The cheek does not blend well with the area around the jaw in both Levels D and E. These characteristics, however, are nowhere near as pronounced as they are in Level A and Level B individuals.

Level D and Level E individuals appear to have changed a little, but are still lacking mentally in comparison to the higher levels of humanity. It has been much remarked upon, that the high level of aggression present in Levels D and E is often displayed. It is a fact that Level D and E individuals lack the subtlety of higher levels, and are unable to display genuine confusion, as well as some other emotions such as empathy and pity. The lips have not yet adapted correctly to display emotion. A distinguishing feature of individuals of Levels A, B, D, E, and C is a certain "quality" about the eyes which is the same "quality" found in the eyes of various lower primates-a kind of moist dullness. While this feature is less apparent in DE than it is in A and B-and begins to drain away at C to almost, but not quite, vanish somewhere around F, G, and possibly H-it nonetheless remains, within DE. The eyes do not display emotion well. The Ainu and Tibetans have been highly discriminated against in the countries in which they live, and Level E individuals have also been discriminated against. They have, however, developed considerably from Level A and B individuals.

Level C

Level C is found among Mongols and also among the Aboriginals of Australia, although Levels K and R are also found among the Aboriginals. The evolutionary distance between Levels B and DE appears to be the same as the evolutionary distance between Levels DE C. However, the gap between Levels C and F appears to be greater.

The language of Australian Aboriginals is significantly more complicated than African languages; indeed many words have been adopted by the Level R individuals of Australia, such as "budgerigar" and "kangaroo".

The history of Aboriginals appears somewhat richer than the African histories. The history of Mongols may be richer still, and the Mongols infamously acquired an empire which covered a significant proportion of the globe.

Voice appears to be somewhat lighter and more toneful than Level DE voices, while hands and feet remain long. The nose is thick, as are the lips, to a certain extent. The ears sometimes protrude.

Voice pattern is still somewhat broken up, although it does appear to naturally mimic the pattern of the higher levels, to a certain extent. On the whole it appears that the voice fluctuates between DE and F, sometimes resembling DE and sometimes resembling F. Walk remains somewhat poorly evolved, although this aspect has improved considerably, and nose formation is still unimpressive. Being an "Asian" level, with eyes appearing "broad", Group C individuals do not have lips and teeth which protrude distinctively, although the length between upper lip and nose is still large. As with all of the lower races of humanity, the cheek area does not blend well with the jaw area, creating a "patch" that resembles the "patch" found in lower primates, lying around the lip area, under the nose, and along the lower jaw and chin.

This level appears to be relatively intelligent. The aggression is still present in quantities, however, and much emotion is still lacking. Signs of real confusion are visible for the first time within humanity, and hints of subtlety also appear. The Aboriginals and Mongols have both been discriminated against, in the past. At this stage the "quality" which is present in the eyes of apes, such as chimpanzees and gorillas, has begun to drain away. In its place has been born a new quality, which defines Levels F, G, and H, but then disappears. This quality is a peculiar "vagueness", and "creepiness" about the eyes, which, instead of making the individual at Level C, F, G, or H seem apelike in any way-although Level C has been said to remain "primate-like" in eye and feature of lip and skull-makes the individual appear gnomish, monstrous, doll-like, or troll-like. This feature is caused by lip, nose, skin, expression, cheek, skull, and eye. It is even a way of defining "larger groups" of levels, although the grouping is controversial, and many individuals will doubtless resent being grouped together. Behold the larger groups: Group A-Levels A, B, DE, and C; Group B-Levels F, G, and H; and Group C-Levels IJK, L, M, NO, P, S, and T. Focus in the eye is entirely lacking in Levels A to H, with the eyes appearing vague.

Level F

Level F individuals are frequently found in Korea. The evolutionary distance between C and F, as stated, appears to be great indeed, while F, G, and H appear similar, reflecting the similarities of A and B, and others.

The voice remains deep and toneless, although much less so than the lower levels of humankind. The hands and the feet are relatively long, although this length is much reduced. While the lips are only of slight thickness, complete thinness of lip may be impossible to achieve at this stage. The ears never appear to protrude, and this aspect shall now be dispensed with in description.

The voice pattern hass now evolved completely, and direct answers to most questions in interviews and the like are now provided. Indeed voice and speech often convince the hearer that this is an individual of a level equal to all levels above. Description of voice shall also now be dispensed with. Walk remains primitive though, although this aspect is now departing from humanity. The lips and the teeth continue to protrude, though only slightly. The skull is not domed, and the forehead does not slope; I shall no longer speak of these when I speak of the levels from this point forth. For "regular" Level F individuals, the lower jaw only protrudes in an insignificant manner, while for Eastern Level Fs the length between the upper lip area and nose still remains large, but this is hardly noticeable. The inability of cheek to blend has now almost gone. This is a Level which, in all likelihood, evolved near northeast Africa, or near Israel. As a result, and in a similar fashion to Levels DE and IJK, the face has been "squashed"; consequently, the eyes are closer together. The diagnostic feature of this level, combined with features of eye, nose, and lower jaw, is the presence of a "beak" in the upper lip-the skin above the upper lip curves downwards at and towards the centre to create a sharp point, which almost resembles a tooth.

This is quite an intelligent level, distinct from Level C. However, on occasion, the individual "switches off", and it has been said that this is doubtless some manner of evolutionary failure. This feature has also been observed in individuals of Level IJK, although this fails to detract from the value of IJK individuals, since Levels IJK, L, M, NO, P, S, and T bear strong similarities. Level F individuals frequently do not "get" this or that. The aggression frequently emerges, although again this differs from its emergence at C. The eyes display emotion-on occasion. The emotion is forced, however, and unnatural, and display is slow; there is a general sluggishness about the display of emotion. The first glimmers of empathy have been observed at F, usually in an attempt to convince the individual with which the F converses, that the F is equal to him or her, although the F is often unsuccessful in the utilisation of this small amount of empathy-the F instead succeeds in appearing even stranger. Nevertheless, a great variety and great number of mixtures of emotion are possessed by Level F individuals.

Level G

Level G individuals can be found near their likely evolutionary location-the area between the Black Sea and the Caspian Sea.

The voice remains deep, toneless, and identical to the voice of Level F individuals. The hands and feet are also similar to the hands and feet of Level Fs, and thickness of nose and lip is also similar to F nose and lip thickness.

Level G walk also bears similarities to Level F walk, as does Level G nose formation to Level F nose formation, Level G protrusion of lip and tooth to Level F protrusion of lip and tooth, Level G protrusion of lower jaw to Level F protrusion of Lower jaw, and inability of Level G cheek to blend to Level F inability of cheek to blend. Levels F and G bear many similarities, and the evolutionary gap between Levels F and G may be very small. G is a "Western" Level, having evolved at a similar location to P. "Western" and "Eastern" levels can in reality be depicted on a scale, with G lying on one end of the scale-having relatively compressed features-and C and O lying on the other end of the scale-having "broader" features.

This is a relatively intelligent level, and it is one which bears similarities to Level F. However, Level G is somewhat more advanced. A curious phenomenon occurs eyesight comes within around six foot of the specimen: the very faint primitive quality, and the overwhelming gnome-like quality, vanish, and the individual appears fully human in look-although if studied it is evident that movements and emotional displays are still underdeveloped. At this level a new manner of emotion is displayed and is present-confidence and belief in self. This awareness is entirely lacking in F. This is the first level which can be described as bearing distinct and obvious similarities to all superior levels.

Level H

Level H can be found in India, doubtlessly having evolved nearby.

H tone of voice, H length of hands and feet, H thickness of nose, and H thickness of lip all resemble their G equivalents.

The walk, the formation of nose, the protrusion of the lips and teeth, the distance between upper lip and nose, and the manner in which the cheek blends or fails to blend, in all ways evokes G.

The mental aspects of this group parallel the mental aspects of G. The gnome-like aspect also persists within this level.

Level IJK

Level IJK evolved near the Middle East. After its evolution, Level I moved into Europe, which resulted in a "spreading" of the IJK feature to result in a "Western" "look". Level J, after its evolution in the Middle East, spread into a large number of areas, resulting in enormous variety in physical form which parallels the variety of physical form of Levels E, O, and R. Level K roughly evolved at the same location as the location its parent evolved at. The evolutionary gap between H and IJK is enormous. Level I is readily found within Europe; Level J can be found in southern Europe, in the area in and around Israel, and in Egypt; and Level K appears only to remain in Papua New Guinea, and among the Aboriginals of Australia.

The history of J is highly impressive, with cultures such as Ancient Egypt, books such as the Bible, and figures such as Albert Einstein, representing intelligence.

The tone of voice is now identical to the tone of voice of the higher levels, as is length of hand and foot, thickness of nose, and thickness of lip. These features shall not be remarked upon henceforth.

Both the walk and nose formation have now advanced, and the lips and teeth do not protrude. The lower jaw does not protrude for IJ, and for K the length of skin between upper lip and nose is not of abnormal length. The cheek blends perfectly.

Mentionings of all these shall be dispensed with. There are, however, manners in which I, J, and K differ from others, but these physical features are location-specific, and would have been present in the higher levels of humanity had they evolved within the vicinity of the Middle East. All individuals at this level possess short necks, giving them a bullish or a hunched appearance. The nose has been remarked upon, and it is an everlasting truth that, combined with the neck, the nose can be used to help identify individuals at this level. As a result of the fact that this level evolved at the location it evolved at, all features are squashed-the eyes are closer together and appear larger, and of a different shape, displaying a different manner of emotion to "Western" groups-a feature shared with Fs, since both having large staring eyes, due to position and form of the eyelids. The amount of eye shown becomes almost spherical, especially near its centre, as opposed to the gradual narrowing through the "Western" form, into the narrow "Eastern" form. The eyes sometimes appear "fixed" in one or more emotions, or types of emotion, and as a result these individuals sometimes have a "worried" look about them. Individuals of this level have a nose which is fully evolved. However, this large "amount" of nose needs to be "squashed" due to the location at which this level evolved, in developmental terms. Hence the nose of individuals at this level always takes on a "large" appearance, and "sticks out" with a distinct groove along the cartilage, often appearing as an afterthought added to the skull, and almost "carrot-like" in appearance, it has been said. This is a distinction of the individuals which evolved at this location. However, it is incredibly difficult from most photographs to distinguish individuals of this level from Level R individuals, with the "Western" Level I individuals representing great difficulty.

This level is of great intelligence. However, aggression persists throughout humanity, and perhaps even T individuals are aggressive in some manner. Individuals at this level are often aggressive. The eyes do not display emotion too well, either, and the lips are not used in the manner the higher levels use them to display emotion. It is frequently the case that individuals at this level do not "get" this or that, and this or that must be explained to them again after it has been said. While they have a strong grasp of morality, how far IJK can go remains to be seen. When individuals of this level speak, they often appear to lack emotion. However, it is often the case that only a thorough analysis of individuals at this level can reveal any flaw, a testament to the extreme closeness of IJK to the highest levels.

Level L

Level L is an Indian level, having evolved near to where Level H evolved.

    This level is highly intelligent, and resembles, in many ways, Level M.

Level M

Level M evolved in a unique location, contrary to all other levels' locations. It may be present in small quantities in India, as well as in Papua New Guinea. It evolved near Papua New Guinea.

The physical features of individuals at this level are shaped in such a way that the individuals constantly appear to display concern, hatred, and seriousness; it is an incredibly hostile-looking level.

This level is highly intelligent. There is, however, much aggression displayed in the eyes, which do not display emotion too well.

Level NO

Level NO evolved near the East, after which Level N spread into the west, while Level O spread east. Level N is present in areas of Europe, while Level O individuals are relatively easy to find in China, Japan, and much of Asia.

The language of Level O individuals is unnecessarily complicated, and not in fact superior to the language of Level R individuals. For instance, the alphabet is made up of an incredibly large number of letters, matched with unnecessarily varied symbols, each symbol taking a large amount of time to "compose" as opposed to easily and smoothly "forming" and "writing" words of the English language. There is absolutely no need for such a large number of letters, either.

The history of individuals at Level O is, understandably, rich. Indeed they have, and are, embarrassing nations which call themselves "Level R nations", although achieving such a task is of questionable difficulty.

This is an "Eastern" level. However, Level N is a "Western" "Eastern" level, as such, which would make Level O "Eastern" "Eastern". Needless to say it was incredibly difficult to find even one individual belonging to Level N. I believe I found one such sample to study however, and confirmed certain aspects of NO in the process. While it is relatively easy to "find" a Level O individual who, through sexual reproduction, has taken on physical features of a Westerner, "finding" a Level N individual is a much more daunting task. I can now state, though, that both individuals belonging to Levels N and O have an upward "kink" at the edge of the lips, and there often has been said to exist an almost "childish" quality to the eyes and lips of the individual, and the word "childish" is the only word which can describe this curious quality found about the NO. Of course, with the features of skull of O individuals, identification occurs due in part to the teeth appearing of a broader and flatter quality; hence when smiling and laughing O individuals appear different to individuals that failed to evolve in the East; the lips move in a different manner to the lips of those who failed to evolve in the East in order to accommodate the teeth and the jaw. The skull has been "flattened" as such. The eyes of both Level N and Level O are mostly dark and cold.

    This level is incredibly intelligent. However it can, with some difficulty, be shown to be at a lower level to P. A testament to the great intelligence of this level is that many of its individuals are fully aware of the factors placing P above them. These individuals seek to eliminate the factors from their own bodies and minds. However, the factors can never be destroyed-only suppressed. As a result, on occasion, there are certain signs that appear, and these signs prove that NO is situated below P. One feature is the unquestioning worship and adoption of almost everything Western in Level O nations by many of its individuals, in contrast to the unquestioning worship and adoption of things Eastern in Level R nations by only a tiny minority of its individuals. This is not a criticism, but merely details a phenomenon which can be called "Westernmania", or perhaps it is a process tied to "Westernization". Another feature is aggression. This never "comes out" in the manner in which it sometimes does at IJK. However, there is always a "challenge" in the eyes-they stare out with hostility and threat. Another feature is the relatively "rushed" nature of individuals, with individuals often rushing into this or that, and appearing out of control in comparison to the steady P. In fact the calmness and coolness of P in comparison to O is readily apparent, both among American Indians and Europeans. Lack of empathy is also present to a small extent, and Level O once or twice fails in "getting" the odd concept. The final feature which distinguishes this level from Level P is lack of emotion present within NO. Pity is an emotion designed to, for instance, help certain of those in need, and certain of those in danger. The purpose of pity is to aid various organisms, various homeless people, and others. It comprises empathy, sadness, and a liking or love for the pitied. While, sometimes with serious consequences, it can be harmful, for example, to indulge in pity, and to become susceptible to pity, to lack pity to a great extent in the manner in which O lacks pity can indeed, on occasion, be a slight flaw. A majority of emotions and emotional mixes are, nevertheless, possessed by NO, and used accordingly.

Level P

Level P, which perhaps should have been designated Level QR, Level PQ, or such like, evolved in Asia, and comprises Levels Q and R. After the formation of Level P, Levels Q and R developed, and spread in their separate directions: Level Q moved east, while Level R moved west. Level Q individuals can be found among the American Indian population, while Level R individuals are readily found among Europeans.

The history of individuals at this level is highly impressive, and requires little introduction; Q is sometimes synonymous with cultures such as the Aztec culture, while R is responsible for a majority of innovations and developments.

Individuals at this level are incredibly intelligent. However, aggressive tendencies still persist, and the eyes always look out with a faint "challenge" in them. This level is also not as sharp as its superiors, with there being a relative dullness about every individual; eye is dead, in comparison. There is a lack of awareness of the external, as this level struggles in comparison with the highly advanced, who can be identified by a high awareness of the external. The eyes of the advanced always follow the speaker during intercourse. Furthermore, multitasking is simple for the superior, and difficult for individuals at this level in comparison. Emotions and reactions are slow, in comparison to rapid change of emotion and fast reflex of levels higher than R, and emotion displayed in eye is often obvious and lacks subtlety, whereas the eye of higher stage frequently displays a multitude of emotions. Both individuals at R and individuals at S and T exhibit a curious feature whereby one eye displays emotion in one form, while the other displays emotion in another form(for instance, one eye may display sadness while the other displays hatred)-a feature perhaps absent somewhat in the lower races. Complete grace is only born within the higher races of humankind.

Level S

Level S in all likelihood evolved at a location near Papua New Guinea, where individuals at this level have been found.

The intelligence of this level is comparable to that of Level T; indeed S and T are closer in evolutionary distance than R and S. However, on occasion a very slight challenge, or sign of aggression, can be displayed by S, distinguishing it from the "challenge-less" T. Both S and T can be characterised by eyes which appear angelic, or sometimes daemonic or alien. The eyes always appear to shine and radiate, and this quality is lacking in R. At these last two stages in human evolutionary development, it is frequently apparent that all emotions are subtle and complex, and sometimes only intense study of photographic, real-life, or video material can identify the various visible mental aspects.

Level T

This is the last level of humanity. It evolved in Asia, and individuals at this level are difficult to locate.

This is the level possessing the greatest intelligence. The "shining" or "glowing" quality about the eyes is omnipresent, and the eyes appear angelic, or sometimes daemonic or alien. Often a negative emotion will flash in the eye, which then transforms into a frightening, ominous, and alien vortex. Reflex is fast, and emotion is complex, often appearing and disappearing about the eye and lip at breathtaking speed. Emotions are frequently of high subtlety. The eyes always follow the speaker during intercourse, and voice is perfect in the conveyance of emotion and meaning. There are doubtless emotions contained within T that the R brain cannot comprehend. Needless to say, many aspects of T remain unresearched.

Conclusion

The physical and mental differences of humanity have now been described; the flow of page must now slow to a halt. All levels have been exposed, but perhaps there is much that remains hidden, waiting to be discovered by the light of revelation, like a candle waiting for its match. While indeed the past is fixed, and the present neither approaches nor departs, the future is, always has been, and perhaps always will be, strange, wondrous, but ever unknown.

www.ingramcontent.com/pod-product-compliance
Ingram Content Group UK Ltd.
Pitfield, Milton Keynes, MK11 3LW, UK
UKHW041433180426
11947UKWH00007B/418